Cooking for Birds

This edition published in 2011
First published in 2006 by New Holland Publishers (UK) Ltd
London • Cape Town • Sydney • Auckland
www.newhollandpublishers.com

Garfield House, 86–88 Edgware Road, London W2 2EA, UK
80 McKenzie Street, Cape Town 8001, South Africa
Unit 1, 66 Gibbes Street, Chatswood, NSW 2067, Australia
218 Lake Road, Northcote, Auckland, New Zealand

ISBN 978 1 78009 068 9

Publisher: Simon Papps
Publishing Manager: Jo Hemmings
Project Editor: Gareth Jones
Editor: Chris Harbard
Designer: Gülen Shevki-Taylor
Artist: Rachel Lockwood
Indexer: Colin Hynson
Production: Joan Woodroffe

Reproduction by Modern Age Repro House Ltd, Hong Kong
Printed and bound by Craft Print International Pte Ltd,
Singapore

Other Books by New Holland include:

Bird Songs and Calls
Hannu Jännes and Owen Roberts. Perfect for the dawn chorus.
Book and CD containing the bird sounds of 96 species.
£9.99 ISBN 978 1 84773 779 3
Also available: *Common Garden Bird Calls*, covering 60 common
species of gardens and parks. £6.99 ISBN 978 1 84773 517 1

Chris Packham's Back Garden Nature Reserve
Packed with practical advice on gardening for wildlife and the
identification of birds, animals and plants.
£12.99 ISBN 978 1 84773 698 7

Colouring Birds
Sally MacLarty. Ideal gift to help develop a child's interest in birds.
Features 40 species, including Robin, Blue Tit and Chaffinch.
£2.99 ISBN 978 184773 526 3
Also available: *Colouring Bugs* (£2.99, ISBN 978 1 84773 525 6).

The Complete Garden Bird Book
Mark Golley and Stephen Moss. A best-seller which explains how
to attract birds to your garden and then identify them.
£9.99 ISBN 978 1 84773 980 3

Kingfisher
David Chandler. Contains more than 80 stunning colour photos
detailing feeding, courtship, nest-building and raising young.
£12.99 ISBN 978 1 84773 524 9
Also available: *Barn Owl* (£12.99, ISBN 978 1 84773 768 7).

New Holland Concise Bird Guide
An ideal first field guide to British birds for children or adults.
Covers more than 250 species in full colour. Endorsed by The
Wildlife Trusts. £4.99 ISBN 978 1 84773 601 7
Other titles in the *Concise Guide* series (all £4.99): *Butterfly and
Moth* (ISBN 978 1 84773 602 4), *Garden Wildlife* (ISBN 978 1
84773 606 2), *Insect* (ISBN 978 1 84773 604 8), *Mushroom*
(ISBN 978 1 84773 785 4), *Seashore Wildlife* (ISBN 978 1
84773 786 1), *Tree* (ISBN 978 1 84773 605 5) and *Wild Flower*
(ISBN 978 1 84773 603 1).

New Holland European Bird Guide
Peter H Barthel. The only truly pocket-sized comprehensive field
guide to the 500 species found in Britain and Europe.
£10.99 ISBN 978 1 84773 110 4

The Urban Birder
David Lindo. The Urban Birder shows how any place, even the
most unpromising cityscape, can be a haven for birds.
£9.99 ISBN 978 1 84773 950 6

**See www.newhollandpublishers.com for special
offers and details of over 200 books on wildlife**

Cooking for Birds

Fun recipes to entice birds to your garden

Mark Golley

New Holland

contents

Introduction

When you browse the titles in your local bookstore, you can almost hear the shelves groaning under the weight of a plethora of cookery titles. Old ways, new ways, Thai, French, Chinese, Indian and Italian. For every Nigella, there's a Delia, each Ramsay is matched by an Oliver and a huge array besides. We seem to love cookery books and cooking for ourselves.

How often though, have you thought about a little kitchen fun and cooking for some rather different guests?

The most frequent visitors that you can entertain with your culinary skill visit every day of the year, come rain or shine. They wait patiently for anything that you care to serve to them, they never complain and you can guarantee that they are always grateful. They may even show their appreciation with a song. No garden should be without them, as they always brighten up a dull day. 'They' are, of course, birds!

Birds are, undoubtedly, the life and colour of any garden, wherever it may be. You can live in the remotest of areas, or in the most populated; your garden can be acres long, or just a few feet wide. But one thing is guaranteed, with food made available, the birds will come.

This book aims to bring you some of the most popular and successful bird food recipes there are. Some are based on childhood successes, others are adapted from popular ideas from sources around the world.

So whether it's a treat for tits, a nibble for a Nuthatch or a bellyful for a Blackbird, there should be plenty here to entice a variety of visitors to your bird table or bird feeders. And, hopefully, the whole family will enjoy making the recipes, and then take pleasure in the reaction of the birds!

The Importance of Feeding

There is no doubting the enormous pleasure that many millions of people across the country gain from feeding the various species of birds they see in their gardens. A survey conducted by the Royal Society for the Protection of Birds towards the end of the 1990s suggested that an astonishing two out of three people put food out for birds regularly during the winter.

There are numerous reasons why we should feed our wild birds. Perhaps the most obvious of all is that, without our help, many of them may die, particularly during severe winters. Smaller species such as Robins, Wrens and Blue Tits are especially at risk because they all use up vital energy searching for food during the short winter days. And so, if we can help them by providing a safe and regular source of energy, then so much the better. Loss of suitable habitat has also become an issue: with the decline of woodlands, hedgerows and weedy field margins, we, in turn, lose plants and insects that are vital components in a bird's food chain, particularly during the summer, when there are millions of baby birds in need of food.

You needn't confine your garden bird feeding to winter and early spring. Over the years, the question of whether to feed birds during the breeding season has been something of a 'hot potato'. Now though, conservation organisations have recognised the benefits and importance of feeding all year round. Providing an easy source of food in the garden during the summer keeps the adults in good condition and leaves them more time to look for the caterpillars, insects and grubs required by the growing chicks, which they will not need to eat themselves!

For the spring and summer it is best to:

- Continue to provide food, but ensure variety at all times (perhaps including live food too).
- Avoid using whole peanuts; a wire feeder will ensure that adults only take small pieces.
- Reduce the amount of food you put out as autumn approaches, as this is the one time of year when natural food supplies should be at their optimum.

In addition to food, you should always ensure that there is a good supply of fresh water available. Many birds need to drink at least a couple of times through the course of a day, so fresh water is invaluable. A separate source for bathing is also a welcome addition for birds in your garden.

For more detailed information on this and all other matters relating to the subject read Stephen Moss and David Cottridge's excellent, information-packed book, *Attracting Birds To Your Garden* (New Holland).

Core Ingredients

Throughout the book several 'core ingredients' are used for a number of recipes. Many of these things are likely to be on hand within the kitchen, but if you feel a *Cooking for Birds* meal-making frenzy coming on, it may be worth ensuring that you have the following essentials available:

Core Ingredients

- Suet (animal or vegetable)

- Wild bird seed mixture

- Peanut butter

- Stale cornflakes or rice crispies (or cornmeal)

- Raisins

- Breadcrumbs

Seeds & Pulses

All manner of seeds and pulses are available throughout the country, in specialist wholefood stores or in most super-markets. The recipes that follow are all incredibly simple to follow and will always prove popular with seed-eating species, such as Chaffinches, Greenfinches, and Siskins. The first recipe doesn't even require any cooking!

Honey Twigs

You may have heard the old joke, 'What's brown and sticky?' 'A stick!'... Well here's the *Cooking for Birds* punchline – a 'Honey Twig'! This recipe takes no time at all to prepare and the seed-eaters that pop in to the garden will be ever more likely to linger with a 'Honey Twig' on offer – they will simply love this sweet-toothed treat! Keep your eyes on those finches and sparrows as squabbles are likely when they begin to taste this oh-so-simple snack. Adapted from treats that are popular with aviculturalists, wild garden birds should really appreciate this easy-to-make recipe.

Ingredients

Several sticks or twigs

Honey (clear 'runny' honey is best)

Wild bird seed

Method

✦ Have a wander in your local park or wood, and search out as many sticks and twigs as you fancy that are an average 1–5cm in diameter. (They can be as long as you want.)

✦ Take a spoonful or two of runny honey and spread it all over the selected twigs and sticks. Now simply pour the wild bird seed mixture over the honeyed sticks. Once they are covered in seeds, place the sticks in the fridge to harden up and then hang them in the garden, from feeders or other twigs.

Bird Balls

This is one of the few recipes in the book that actually has a specific

measurement! For those with an aversion to metric measurement,

a pound of lard will do for this wonderfully messy recipe.

Ingredients

½kg (1lb) lard

1 jar crunchy peanut butter

5 cups cornmeal (available in many wholefood shops) –
crushed stale cornflakes/rice crispies make a good substitute

6 cups quick-cook porridge oats

2 cups sunflower seeds

2 cups raisins

Method

🪶 Cut the lard into smallish blocks so that you can soften it sufficiently when mixing in other ingredients. Once soft enough, put the lard blocks into an old mixing bowl and add the peanut butter, cornmeal and porridge oats. Using your hands, mix everything together. When you are happy that all the ingredients are well mixed, divide the stodgy mess into half a dozen 'portions' and roll them into ball shapes. Finally, roll these balls into the sunflower seeds and raisins.

🪶 Suspend the balls from your bird feeders inside some sturdy netting, or place the balls inside a bird feeder. Alternatively, form the balls around sturdy twine and tie them to nearby branches. Then it's time to wash up!

Nice 'n' Seedzy

As well as slaving over a hot stove for your garden bird visitors, it's nice to be able to offer them up something which takes no time at all to prepare, involves barely a bowl being offered up to the dishwasher and is incredibly nutritious too. All you need are seeds. The finches that pay you regular visits will love this simple mixture. And, as mentioned in the Method (see opposite), you could try adapting the recipe to make seed balls too... You could also add some crushed cream crackers into the mixture as an additional nibbly treat.

Ingredients

2½ cups sunflower seeds

1 cup millet seeds

½ cup wild bird seed

Method

✤ Pop all the seeds into a bowl, mix, and serve in a seed hopper/feeder! Voila! A guaranteed hit. You can also bind the seeds together and form seed balls using suet (animal or vegetable suet is fine) and perhaps some peanut butter. A combination of both works particularly well, but melt them both down before adding the seeds. As the mixture cools, pop it into the fridge or freezer compartment to ensure the fat hardens properly.

Summer Sunflower Seedheads

This is another seeds 'recipe' which needs very little attention to detail! In fact, seeing as you can leave the kitchen out of this altogether, it's a bit cheeky to even call this a recipe! This idea should prove really popular with the finches that pay you regular visits.

Ingredients

Sunflower heads

Method

🌾 Many people are keen gardeners and some may also grow sunflowers. What is needed for this easy bird feeding idea is a number of seeding sunflower heads. If you don't grow your own, take a wander around the countryside. Many farmers grow sunflowers as cover for gamebirds (that they shoot), and may let you have a sunflower head or two. Alternatively, have a chat with a friendly florist, or even a local supermarket which sells flowers.

🌾 Cut off the sunflower heads, along with around 30–36cm (12–14") of stalk. Take a piece of strong twine (rope or tough string will suffice) and tie it securely to the stalk (you may also need to tie some twine around the flower head too). Then secure the flower head to the side a bird table, a feeding station pole or even to the branch of a tree.

Seed Cornbread

This is an American recipe that is especially popular with garden birds in the States. If you can get hold of cornbread mixture (it is available in the UK), give this simple, quick and easy recipe a try. The results are amazing, as Blackbirds, Song Thrushes, House Sparrows and even Crows will scrimmage around the bird table for a taste.

Ingredients

1 packet cornbread mixture

1 extra egg (i.e. one extra to the amount required for the cornbread mixture)

1 cup wild bird seed

Method

✎ Make up the cornbread mixture as per the instructions on the packet. Once the mixture is made, add the extra egg (shell included), along with the wild bird seed (feel free to vary the amount, as required). If the mixture is a little dry after adding the shell and seeds, just add a drop more water. Then bake as per the cornbread packet instructions.

✎ Once baked, allow the bread to cool before cutting into reasonably sized pieces. Freeze any pieces that you don't use and defrost them when required. Crumble the slices before putting them on to your bird table.

Beans & Rice

Although this is one of the more time-consuming recipes featured in *Cooking for Birds*, it is extremely rewarding. Full of goodness and nutritionally sound, 'Beans & Rice' should be a wow with thrushes and Robins, especially in wintertime. The beans and lentils used here are widely available in many wholefood stores and supermarkets.

Ingredients

¼ cup pinto beans	¼ cup lentils
¼ cup kidney beans	1 large chopped carrot
¼ cup butter beans	1 cup cooked brown rice
¼ cup haricot beans	1 cup cooked white rice
¼ cup split peas	1 cup sweetcorn

Method

✦ Soak all of the beans, peas and lentils in a pan overnight before cooking. When you have done this, drain them and rinse with cold water. Then refill the pan with water before adding the carrot. Slowly cook the beans and carrot on a low heat for around an hour.

✦ Add the pre-cooked rice, along with the sweetcorn. Cook the mixture until the rice and corn is warmed through. Allow the 'Beans & Rice' mixture to cool down thoroughly before putting it out onto the bird table. You may choose to put the mixture into a container before putting it outside.

✦ Any 'Beans & Rice' mixture that is left can be put into the freezer – just remember to thaw it out before giving it to the birds!

Bean Soup

A variation on the previous recipe that also proves popular with a number of garden bird species: bursting with protein and just as easy to prepare as 'Beans & Rice', 'Bean Soup' ought to be given an avian nod of approval. Thrushes will be more than happy to step up to the bird table for some of this, and try getting the Starlings to stay away! As mentioned in the recipe itself, there are plenty of alternatives you could consider using, so why not experiment and come up with your own version.

Ingredients

1 bag vegetable and pasta mixture (available in most supermarkets and contains peas, beans and pearl barley etc.). As an alternative, use as many varieties of pulses as your cupboard can run to.

1 cup uncooked brown rice

1 bag frozen sweetcorn

Method

★ Follow the instructions on the packet – a vegetable and pasta mixture usually takes around 50 minutes to cook through. If you are choosing to use your own pulses then you will need to soak them overnight.

★ If using the vegetable mixture, add the rice as soon as the water has come to the boil, and when the 50 minutes or so are up (or the pulses have begun to soften), add the frozen sweetcorn. Once the sweetcorn has cooked through, drain off the excess water and allow the 'Bean Soup' to cool.

★ Once cooled down, place in a tub or container and take out to the bird table. Any that is left over can be frozen.

Millet & Rice

This recipe is something of a 'winter warmer'. While it can be served up

at almost any time of the year, it's best to avoid the middle of summer.

Seed-eaters such as finches will enjoy this dish, as will thrushes,

especially if you add a little chopped apple to the mixture.

Ingredients

½ cup cooked de-husked millet (sweetcorn can be used as an alternative)

½ cup cooked white rice

½ cup cooked brown rice

Method

✎ Soak the millet in water for around 30 minutes before you start to cook the mixture. Prepare the white and brown rice together as you would normally, and gently warm the millet through until soft.

✎ Once you have cooked the millet and the two rices, mix them together. Then, while the mixture is still warm (but not hot) pop the 'Millet & Rice' out onto the bird table.

Popcorn Strings

Another winter favourite that is wonderfully simple to make. Almost every bird that visits the garden, from Great Tits to Dunnocks, Nuthatches to Redwings, will find something tasty and nutritious for them on the end of 'Popcorn Strings'.

Ingredients

1 packet uncooked popcorn

1 jar honey (sunflower or olive oil will do as an alternative)

1 cup wild bird seed

1 cup breadcrumbs

Method

🖋 The joy of this recipe is the simplicity of it all. Simply cook your popcorn as per the instructions on the box. Once the popcorn is ready, head towards the sewing box! Take a needle and some strong thread. Thread the needle, and begin to pass the thread through the freshly-cooked popcorn. Make the 'Popcorn Strings' as long as you fancy and, if you wish, simply hang them up around the garden. As an alternative, you may want to add a little more colour and flavour to them first: brush on some runny honey or some oil, and then roll the strings of popcorn in wild bird seed or breadcrumbs. You may want to chill the strings after that or hang them straight out in the garden.

🖋 One final, colourful, addition to the strings would be some soft fruit. If you have raspberries or blackberries in the freezer from the autumn, defrost a few and alternate fruit–popcorn–fruit–popcorn on your strings. Failing that, try some fresh cranberries – your local thrushes will be so grateful!

Breads & Buns, Pancakes & Peanut Butter

One of the staples of the human diet is bread. It's almost as popular with garden birds! While many of them enjoy nibbling on any breadcrumbs that you leave out for them, they are sure to enjoy the following recipes even more! As well as bread recipes, I've also included some recipes that use another bird recipe staple, peanut butter. This gives the birds a tremendous energy boost and the recipes are easy to do and fun too!

The Fabulous Flying Bird Bread I

This recipe and the next are adapted from highly successful ideas that have been a real boon for garden birds in America for decades. The combination of ingredients will prove popular with many different species and provides perfect nourishment for many garden visitors throughout the year.

Ingredients

2 cups baking mixture (any cake or pancake mixture will be sufficient)

2 cups white or yellow cornmeal (or crushed stale cornflakes/rice crispies)

1 cup wholemeal flour

2 tablespoons baking powder

4 medium or 3 large eggs

3 small jars fruit purée or vegetable meals (the sort that are suitable for small babies)

Method

✒ Pre-heat your oven to around 190°C, 375°F (Gas Mark 5). As the oven is warming up, grease a large cake tin, around 23 x 30cm (9 x 12").

✒ Combine all of the dry ingredients in a bowl and then add the eggs and the jars of fruit purée and vegetables. Stir the mixture well until all the dry ingredients are nicely moistened. By this time the mixture will be very heavy and thick. Now pour the mixture into the cake tin.

✒ Bake the bread for around 40 minutes or so, or until the top has become golden brown and fairly hard. Once the loaf is cooked, allow it to cool completely before you cut it into cubes ready for the bird table or your bird feeders. Keep enough cubes aside for a few days' feeding and freeze the remainder to use as and when you please.

✒ The recipe can be modified to include almost anything that you fancy using. Anything from chopped fruit and vegetables to peanut butter and cheese will work perfectly well within the 'Fabulous Flying Bird Bread I'.

The Fabulous Flying Bird Bread II

Another great bird bread recipe and, again, one that is pretty simple to put together. Try to find some packets of cornbread mixture – they may be tricky to find but work really well with this recipe. The blend of ingredients will prove popular with many species that visit the garden.

Ingredients

2 packets cornbread mixture

1 extra egg (i.e. one extra to the amount needed for the cornbread)

Fruit juice of your choice

½ cup de-husked raw sunflower seeds (available in supermarkets, as well as wholefood shops)

1 cup fresh, chopped vegetable of your choice

1 cup fresh, chopped fruit of your choice

½ cup raisins

½ cup de-husked raw millet seed (available in wholefood shops)

Method

✦ Pre-heat your oven to around 190°C, 375°F (Gas Mark 5). As the oven is warming up, grease a large cake tin, around 23 x 30cm (9 x 12").

✦ Make up the cornbread mixture as per the packet instructions, substituting the amount of milk specified in the instructions with fruit juice, and adding the extra egg.

✦ Once the eggs and fruit juice are combined, add in the sunflower seeds, the vegetable and the fruit. Mix together well before adding the raisins. Mix well, then pour into the cake tin. Before you put it into the oven, sprinkle with the millet seed.

✦ Bake the bread until golden brown and hard on the crust. Once the loaf has cooled, cut it into cubes and place some crumbled cubes onto the bird table, or into a bird food holder. Any leftovers can be frozen and used as you wish.

✦ You can use cooked pasta or beans instead of vegetables. Peanut butter or apple sauce can also be used (if you use either, reduce the liquid accordingly).

The King's Bread Pudding

One of my very favourite recipes to make, and one that the birds seem to love, this is a winter recipe, named in tribute to the similarly fearsome snack of choice that 'The King' enjoyed so much in his later years. This is crammed full of flavoursome items that all manner of garden birds will keep coming back to!

Ingredients

1 large, fairly stale, unsliced bread loaf

1 jar peanut butter

1 cup peanuts

½ cup sunflower seeds

1 cup raisins

½ cup diced apple

½ cup sweetcorn

Method

✒ Cut the crust off one end of the bread (break it up and pop out onto the bird table). Then, with your hands, or a spoon, hollow out the inside of the loaf as much as you can. (Scatter these breadcrumbs onto the bird table too).

✒ Once the bread has been hollowed out, use a skewer to make two holes in the other end of the loaf, through the end crust, keeping the holes as far apart as possible. Thread some strong string or twine (even thin rope) through the two holes.

✒ Start to fill the loaf cavity with a mixture of peanut butter, peanuts, seeds, apple and sweetcorn, making sure that the peanut butter is the ingredient that you use most of. Once the loaf is full to the brim with ingredients, you can hang it up in the garden. You may want to experiment with 'perches' for the birds within the loaf. Funnily enough, cream crackers on the top of the bread may work, but they aren't so successful on the side. It's a case of trial and error to find what works best!

Peanut Butter Balls

As with lots of other recipes in the book, this one is quick and easy to make and the birds really do seem to enjoy the combination of peanut butter, cereal, nuts and fruit. If it wasn't for the lard, you might be tempted to have one too! Your local Blue Tits and Great Tits will enjoy them, and if they don't devour them all, maybe a Great Spotted Woodpecker will sneak in for a nibble.

Ingredients

1½ cups lard

1½ cups crunchy peanut butter

5 cups crushed stale cornflakes/rice crispies (or cornmeal)

1 cup chopped, mixed nuts

½ cup chopped dried fruit

Method

🔸 Melt the lard slowly over a gentle heat. Once the lard has melted completely, start to add in the peanut butter, stirring until it has combined well with the lard. Add in the five cups of crushed cornflakes/rice crispies and, once the cereal has been absorbed into the mixture, drop in the in the mixed nuts and finally the dried fruit. If the mixture is still a bit runny, add in some more cornflakes/rice crispies to stiffen things up a little.

🔸 Allow the mixture to harden in the fridge. You can then put it into a sturdy onion bag or, preferably, into a square mesh feeder.

Rock Cakes

This is such a simple garden bird snack to make! Almost identical to the traditional just-like-granny-used-to-make recipe, little effort is required to create the finished product. A nice, crumbly, fruit-laden cake appears at the end of the cooking process and will be devoured by Robins, Starlings, House Sparrows and more besides. These 'Rock Cakes' can be put out directly onto a bird table, or suspended in a wire mesh holder.

Ingredients

1 large size tub margarine – around 225g (½lb)

1kg (2lb 2oz) self-raising flour

A little water

4 tbsp sugar

2–3 generous handfuls raisins and currants

Method

🕊 Pre-heat your oven to a moderate temperature – around 180°C, 350°F (Gas Mark 4). As the oven warms up, grease a large cake tin, or a couple of bun trays, ready for the simple mixture.

🕊 Before combining the ingredients, make sure that the margarine is softened first as (just like cooking for yourself) this will help speed up the process when blending the ingredients. Once the margarine is nice and soft, add the flour (sift it if you have the time). Add a little water at this point to help to thicken the mixture. Next add the sugar and fruit, mix well and again add a little water to thicken the mixture if required.

🕊 Pour the mixture into your cake tin or bun trays and bake until the crust of the 'Rock Cake' is golden brown and firm to the touch. Once cooled, pop out as much as you want onto the bird table. The rest can be frozen for another time.

Pancake Palava

This is not one of those rather puny pancakes that we struggle to toss properly every Shrove Tuesday. No, this is a recipe made from those large, fluffy pancakes that are one of the staples of the American breakfast. Luckily for us (and now the birds!) the mixture that makes these delicious pancakes is widely available here. This recipe is almost as quick and easy to prepare as the pancakes themselves. Your garden visitors will be most grateful when you just can't manage to squeeze down a fourth one!

Ingredients

1 box American-style pancake mixture

1 cup chopped, mixed nuts

1 cup raisins

1 cup chopped, dried fruit

Method

✒ Assuming you do manage to squeeze that extra pancake in, then why not make the birds their very own batch? Follow the instructions on the packet, make up the pancake mixture as usual, and cook them as you normally would.

✒ Once the pancakes are in the pan and the batter still cooking, add in a handful or two of your chosen ingredient, be it nuts or fruit. Continue to cook the pancakes as normal, turning once.

✒ Allow the pancakes to cool down, then tear them up and pop them outside and onto the bird table.

✒ Any leftover pancakes can be stored in the freezer, but remember to place a sheet of greaseproof paper between them to help separate them.

Pancake Bird Bars

As if tempting you with one pancake recipe isn't enough, here comes another one! Actually the end product is far less appetising to our eyes – more a bird bar than a pancake – but as pancake mixture is the core ingredient it's allowed the grand 'Pancake Bird Bars' title! Almost every species in the garden will take a nibble at these wholesome snacks.

Ingredients

2 eggs

1 cup wild bird seed

1 large carrot

½ cup fresh, raw broccoli

⅓ cup mixed nuts (not peanuts)

2 boxes American-style pancake mixture

1 jar potato baby food (sweet potato would also work well)

1 jar green bean baby food (another green vegetable would be suitable)

Method

✦ Pre-heat your oven to around 190°C, 375°F (Gas Mark 5) and grease a large baking tray. As the oven is heating, mix the following ingredients in a blender: the two eggs (including shells), the wild bird seed, the vegetables and nuts. Whizz them through at high speed until everything is well and truly mashed and puréed.

✦ Drop the pancake mixture into a large mixing bowl. Pour in the purée of egg, vegetables and nuts, followed by the baby food. With a sizeable spoon, stir the mixture until everything is blended well (this may take several minutes) and is fairly smooth in consistency (though it should remain sticky and a bit lumpy). If it seems to be a little on the thick side, add a little water.

✦ Pour or spoon the mixture into the baking tray and put it into the oven. Bake for around 30 minutes, until golden brown. Ensure the mixture is cooked through.

✦ Allow the cake to cool for a few minutes, but while still warm, cut it into bars. Put the bars out immediately, freezing any that are left over for another time.

Breakfast Bagels

It's not just people who enjoy taking their place at the dining table to nibble on a breakfast bagel. Think of all those species that squabble over the breadcrumbs that you put for them – the Starlings, Blackbirds and Robins. All of these species and a few more besides will enjoy this incredibly quick and easy bready treat.

Ingredients

1 pack plain bagels

1 jar honey

Several generous handfuls wild bird seed

Method

✦ Prepare the bagels as you would if you were warming them through for yourself. Once they are ready, split the bagels open and pour or spread on the honey.

✦ Once sufficient honey is oozing into the bagel, simply cover each half with wild bird seed (the honey should ensure that a lot sticks) and pop them out into suspended wire mesh feeders.

Bird Bread Pudding

Bird food recipes are, often as not, successful variations on a theme. This recipe certainly follows in the footsteps of other bread pudding recipes that have come and gone in the cooking for birds history books. Simple to do and undoubtedly a big hit with many garden bird visitors.

Ingredients

Approx. 800g (½lb) lard or margarine

3 good cupfuls porridge oats

1 large cup currants (or raisins)

Good handful grated cheese

Good handful wild bird seed

Method

🪶 Gently melt down the lard or margarine over a medium heat. Once the fat has melted, gradually mix in the oats, followed by the currants, cheese and the wild bird seed. Combine the ingredients well, then spoon out into any suitable containers that you have spare (margarine tubs are ideal). Alternatively, form them into shapes that will fit into your sturdy wire mesh feeders.

🪶 Once the 'Bird Bread Pudding' mixture is in place in containers, pop them into the fridge, allowing them to set and firm up before serving them up on the bird table.

Fruit & Veg

Many birds love to feed on fruit. Barely an autumn or winter goes by when windfall apples aren't being devoured by local Blackbirds, or Redwings and Fieldfares from more northerly climes. As well as cutting up various fruits – from apples to oranges, plums to nectarines – try some of your local birds out on some raw veggies too. Broccoli, tomatoes and green peppers appear to be popular here. Do be careful in warm, summer weather – try to ensure that the fruit and veg stays mould-free and doesn't spoil.

Fruity Ice

Here's a slightly wacky recipe, for the summertime. See how the garden bird visitors fare with this. Stunningly simple, the thrushes should enjoy this, and it is doubtful that any other species would get a look in!

Ingredients

As many types of fresh fruit as are available

Method

- Take all your fruits and dice them into medium-sized chunks. Pop them into a food processor and mash them until smooth and thoroughly mixed.

- Once the fruit is nicely puréed, put the liquid into ice cube trays (or ice cube bags if you prefer) and freeze until totally solid.

- Use the cubes as and when you think the birds will fancy them. They should prove to be a funky summertime alternative!

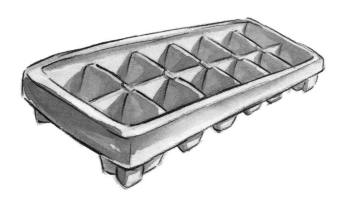

Lollipop, Lollipop!

It has long been known that cagebirds have something of a penchant for nibbling, chewing and licking sweet treats, but this is also the case for wild birds too! Here's a recipe, similar to 'Honey Twigs' but with a little more on offer, for the more discerning garden bird visitor!

Ingredients

Lollipop sticks

½ cup mixed, dried fruit

½ cup wild bird seed

½ cup chopped nuts

1 egg

Some 'runny' honey

Method

🪶 Arrange your lollipop sticks on a baking tray and pop them in to the oven, warming them at around 35°C, 100°F (below Gas Mark 1).

🪶 As the sticks are gently warming through, put the nuts, fruits and bird seed into a mixing bowl. Crack the egg (discarding the shells) and bind all the ingredients together, until everything is nice and sticky!

🪶 Remove the heated lollipop sticks from the oven and turn the heat up to around 70°C, 200°F (Gas Mark 1). While the sticks are still warm (but not too hot), start to form the lollipops. Grab a small handful of the mixture and form it into a ball around the stick. You can, of course, vary the size of the lollipops.

🪶 Put the newly formed lollipops back into the hot oven on the baking tray and bake for 20–30 minutes. Once they appear nicely toasted, then they are done.

🪶 Take the toasted lollipops from the oven and coat the whole of the lollipop and the stick with honey. Then put them back into the oven for around 5 minutes.

🪶 Allow the lollipops to cool completely before putting them out into the garden. You can hang them up or insert them into bird feeders that may already be hanging in the garden. Any leftover lollipops can be stored in an airtight container for several days.

Funky Alfresco Fruit Salad

Here's another summertime recipe to give the hard working bird parents a little treat. No cooking is required and you can prepare this colourful and nutritious dish in a matter of minutes.

Ingredients

2 medium-sized oranges

Small bunch seedless grapes (around 30 or so)

1 small banana

1 small apple

1 cup strawberries

Method

✒ Prepare your fruit. Peel the oranges, peel and slice the banana and cut the strawberries in half. You may want to cut the grapes in half too. Cut and dice the apple.

✒ Once all the fruits are cut and sliced, all that remains to be done is to combine all ingredients in a bowl and mix them well. And that's it! Put the fruit mixture into a bird bowl and out on to the bird table. Any remaining fruit salad can be frozen.

Corn-on-the-cob – With a Twist!

The best thing with any recipe, whether cooking is involved or not, is simplicity.

Many of the recipes within the book are deliberately simple, and this idea is

as simple as they come. All you need are some corn-on-the-cobs and some fruit.

Hopefully, within no time at all, a variety of species, from Robins to Dunnocks,

House Sparrows to Blackcaps, will be investigating this colourful meal.

Ingredients

1 fresh ear corn-on-the-cob (still in its husk)

Small selection of seasonal fruit

Skewers, corks (to ensure that no sharp skewer points injure the birds),
and strong string or twine

Method

✒ Carefully remove the husk and the silk-like strings from the corn. Gently push the skewer through the cob and then, at either end, pop on some chunks of fruit, prior to hanging. Using the twine, tie the fruit-laden corn cob onto an overhanging branch or onto a sturdy feeding station.

✒ Alternatively, you can keep the fruit and the vegetable separately, and hang the corn on its own and use the skewers solely for fruit.

Fruitcake-come-cupcake

Everyone loves fruitcake, even the garden birds. Sparrows chivvy the Robins who duff up the Dunnocks, as they all try and get a slice of the action when stale, dry, fruitcake is put out on to the bird table. But instead of using the stale remnants, how about spending some time preparing the birds their very own 'Fruitcake-come-cupcake'?

Ingredients

2–3 sweet potatoes (or jars of vegetable baby food)

Summer fruits of your choice (fruit or fruit-based jars of baby food will do)

3 eggs

1½ cups wild bird seed

Method

* Prepare and cook your sweet potatoes as you would normally. For speed, you may choose to cook them for around 20 minutes or so, until they are soft enough to mash.

* Once the sweet potatoes are soft enough to mash, pop them into a food processor and purée them – there shouldn't be any need to add any liquid to this purée. Then put the fruit into the food processor and blend together. Pre-heat the oven to 200°C, 400°F (Gas Mark 6).

* Add the eggs (including shells) to the sweet potato and fruit purée. Whizz the food processor again, until the shells are totally blended into the mixture. Once the eggs have been blended, add in the wild bird seed and blend again for a few seconds, until you have a smooth mixture.

* Grease a bun tray and then pour around a quarter of a cup of the mixture into the individual compartments on the tray. Place the tray into the oven and cook for around 15–20 minutes, or until the 'Fruitcake-come-cupcakes' are firm to the touch.

* Once cooked, allow the cakes to cool down completely before putting them out for the birds.

Blueberry Butter Balls

Quick and easy, here is another 'no need for the oven' recipe, guaranteed to be a draw to those species that love both fruit and seeds… and peanut butter!

Ingredients

7 cups plain cooked popcorn	1 jar peanut butter
Handful peanuts (out of their skins)	1 cup raisins
Handful sunflower seeds (in husks)	1 cup blueberries
Handful crushed cornflakes/rice crispies	1 cup blackberries

Method

✒ Cook the popcorn, ensuring you have enough to fill around seven cups. After that, let the mixing begin!

✒ Put all the ingredients into a large bowl, apart from the berries and combine with the gooey peanut butter and the popcorn, until everything is well coated. At the last minute add in the berries. Try and keep the soft fruit as whole as you can.

✒ Put the mixture into some tough mesh bags and chill them (to help them set) before hanging them out in the garden.

The Pleasure of Pasta!

The great thing about this recipe is that you don't have to cook the core ingredient especially for the occasion. All you need to do is ensure that, whenever you cook your spaghetti, penne, rigatoni or macaroni, you make extra! Given the rather starchy nature of the main ingredient, this is perhaps one of those winter-only recipes, giving the birds a little carbohydrate boost on a chilly day.

Ingredients

Sunflower, vegetable or olive oil

Raisins

Pasta (dried or fresh)

Fruit of your choice (fresh or tinned)

Method

✦ Prepare your pasta as per the cooking instructions on the packet.
If you are using spaghetti, try to remember to break it into smaller
pieces before cooking.

✦ While the pasta is cooking, grab a couple of handfuls of raisins and soak
them in warm water so that they 'puff up' a little. Drain them after a few
minutes and they should look suitably juicy.

✦ Once the pasta is cooked, drain and then add sufficient oil to thinly coat
the pasta. Add in the raisins and any fruits you wish to include –
apples are always a good option. Remember to cut the fruit
into small pieces and to drain any juice from tinned fruit.

✦ Allow the mixture to cool, then pop it onto the bird table or into a
suitable sturdy feeder.

✦ As an alternative, you can use peanut butter instead of oil, with raisins
and fruits as an optional extra.

Super Squidgy Suet

Many birds love suet, but remember that suet (and lard) are both animal products, and those people with a love of birds who are vegetarian might prefer not to have to handle animal products. As an excellent substitute, and recipe alternative, you can use vegetable suet or vegetable shortening, both of which are easily found in almost every supermarket these days.

A Few Words About Suet...

Before going any further, it might be useful to explain what is meant by 'suet' and 'lard'. In terms of feeding birds, the word 'suet' has come to mean any generally fatty meal, but there is actually a specific fat type called suet – this being the fat found around a cow's or lamb's kidney. The word 'lard' used to apply just to pork fat, but is used in bird food recipe terms to refer to any animal fat.

Vegetable shortening (and suet) was first launched in America as long ago as 1912. Popular in the home because it lasted far longer than animal fat, vegetable shortening is, basically, 'fats' from vegetables, which can be used in cooking. Vegetable shortening lasts far longer in warm conditions than any kind of animal fat, which should be borne in mind when feeding garden birds.

Many species are very fond of feeding on food items that have suet as a major ingredient. Woodpeckers and Nuthatches are species that thrive on suety recipes, but other species, such as Blackbirds, Song Thrushes, Great Tits, Blue Tits and Robins will also be more than happy to feed on suet-based foods. The energy source that suet provides is invaluable to birds, particularly during winter months and spells of cold weather. Many species burn energy at an amazing rate, particularly in very cold conditions, or when they are especially active in the build-up to the breeding season.

To help birds as best you can, try to ensure that the species visiting your garden can feed on something that is both high in energy (i.e. high in fat) and easy to digest. Note that because of its chemical composition, lard is about twice as hard to digest as vegetable shortening and birds will gain more energy out of vegetable shortening than they will from animal fat. Vegetable shortening and vegetable suet are also healthier for garden birds because they contain around half the saturated fatty acids of animal suet and lard. The vegetable-based fats also contain more vitamins and essential nutrients while lard has only trace amounts of them.

However, this should not detract from the use animal fats have in the feeding of garden birds. A visit to the local butcher for some real suet of the beef kidney type will provide birds with a massive energy boost from its high fat and protein content. Given the problems of animal fats going off in warm weather, it may be best to use this as a treat only in winter.

Within this section, a number of the recipes featured make use of the, now-familiar, square mesh plastic-coated metal cages, which are readily found in shops that sell wild bird foodstuffs. As an alternative, you can try using something like a red onion bag, but these really aren't up to the job, as they tear easily and perhaps last just one or two visits.

Simply Suet

This is a very simple, very quick recipe that should see instant results when hung in a feeder or container. Almost every species of garden bird will find something in the recipe to excite them!

Ingredients

1½ cups melted suet (animal or vegetable)

1½ cups brown breadcrumbs (white breadcrumbs are fine too)

¼ cup granulated sugar

2 tablespoons crunchy peanut butter

1 cup porridge oats

½ cup plain flour

½ cup wild bird seed

Method

🖊 After melting the suet, simply mix all of the ingredients together in a large saucepan, over a fairly low heat. Keep stirring until all the ingredients are thoroughly mixed together.

🖊 When you are happy that everything is blended well, start to spoon, or pour, the mixture into your chosen containers (sturdy paper cups are really good for this) and let them cool down until hard. Before spooning the mixture in, make a hole in the middle of the base of the cup and thread a piece of strong string through, with a knot in the end of it and pour the mixture around the knotted string.

🖊 Once nice and hard, take the cup away, leaving the string in place to tie around small branches. Alternatively place slabs of the mixture into your feeders.

Simply Suet – A Veggie Version

This recipe uses only vegetable suet/shortening as one of its main ingredients. Many of the suet recipes included in the book are suitable for using either animal or vegetable suets. Obviously, if you wish to, you can use animal fat here, but the birds seem to like it as it is.

Ingredients

1 cup vegetable suet or shortening

1 cup peanut butter

½ cup shelled peanuts

½ cup raisins

½ cup cornmeal (or cornflakes/rice crispies)

½ cup maize/sweetcorn/cracked corn (use whichever you can find!)

Method

✎ Take the vegetable suet and melt it down in a large saucepan over a gentle heat. Once this is done, add in the peanut butter and allow that to melt into the suet mixture. Once the two are blended together, slowly add in the remaining ingredients, one at a time. Mix them together well until everything has stuck together (yes, it is that simple!). Spoon the ingredients into a container (a baking tray, or paper cups with added string for tying to a branch) and allow them to harden. You may wish to chill, or even freeze, them briefly.

✎ You can add anything to this (and many other recipes), with small apple chunks or wild bird seed always proving extra popular.

Woodpecker Goo

A true classic! With a name shamelessly borrowed from an American recipe (it sounded better than 'Woody's Treats'!), this is one that has been used for decades in the keen garden birdwatchers' cookbook, and was the very first one I made when I was in short trousers!

Nothing matches the excitement of a woodpecker arriving to feed in the garden. Flashes of black and white zip by, and there, in all its glory, is a 'Great Spot'. There's no better way to lure one of our most colourful and characterful birds than with a log-full of 'Woodpecker Goo'.

Ingredients

Logs (as few or as many as you'd like – no more than 60cm (2') in length)

Suet (animal only)

Bacon scraps (optional)

Method

✦ Take a log and drill up to eight holes in it each around 2½cm (1") in diameter and around 4cm (1½") deep. You can also add a small perch beside selected holes while the woodwork tools are to hand.

✦ Push the suet into the holes. You may be able to get suet with small amounts of meat remaining within the fat. If you can, so much the better, as many birds enjoy them. If not, you could always add small bacon scraps.

✦ Some people prefer to render the suet and then pop the soft suet mixture into the holes once it is cool. If you do render the suet, don't take out any meat scraps. You can subsitute peanut butter for the suet.

Chewy Bird Bars!

Well these aren't really chewy bars, but they certainly resemble the chewy

cereal-based bars that are available in most shops. The recipe comes

in two stages, but is certain to please many species, particularly

Chaffinches and, if you are lucky, Bramblings.

Ingredients

½ cup suet (animal or vegetable)

½ cup peanut butter (crunchy is best)

2½ cups cornmeal (or crushed
stale cornflakes/rice crispies)

1 cup wild bird seed

Additional suet

¼ cup of peanuts

½ cup raisins

1 apple, chopped

Method

✦ Melt down the suet, then add the peanut butter, cornmeal/cornflakes/rice crispies and seeds. Mix thoroughly on a low to moderate heat and press out onto an old baking tray. Place the tray into the freezer compartment until the 'Chewy Bird Bar' mixture is firm enough for you to crumble. When it reaches this stage, crumble the whole batch and place it into a large mixing bowl.

✦ Add more suet, along with the peanuts, raisins and chopped apple. If you feel that more fat is needed for binding the mixture, then add some. There's no exact science to this recipe!

✦ When the ingredients are well mixed for the second time, fill your containers (the mesh feeding 'cage' or paper cups are fine) and freeze them until required.

Chunky Seed Balls

Both the following two recipes have proved very popular with the
visitors to many gardens. This first recipe is an adaptation of
'Chewy Bird Bars', without the addition of fruit, something
that some species appear to appreciate.

Ingredients

3 cups suet (animal or vegetable)

1½ cups crunchy peanut butter

1½ cups crushed stale cornflakes/
rice crispies (or cornmeal, if possible)

3–4 cups wild bird seed

Method

✦ Melt the suet gently in a large pan over a low heat. Once the suet has turned to liquid, add the peanut butter, cornflakes/rice crispies/cornmeal and bird seeds. When the ingredients have been mixed well, turn the mixture out into an old baking tray or cake tin (cutting into usable chunks) and freeze them until required.

✦ As an alternative when using animal suet, you may wish to render it before mixing with the other ingredients. To do this, melt it over a very low heat and allow to cool before repeating the process. While in its melted state, add the additional ingredients.

Sweet Peanut Treats

Another devilishly quick and simple recipe for a popular bird snack.

The high fat content of the lard will appeal to woodpeckers especially, while

the sweetness provided by the sugar will prove popular with thrushes,

finches and tits.

Ingredients

1 cup lard (no substitutions allowed here!)

1 cup crunchy peanut butter

2 cups porridge oats (the quick-cooking microwave type)

2 cups cornmeal (or crushed stale cornflakes/rice crispies)

1 cup plain flour

¼ cup granulated sugar

Method

✦ Melt the lard and the peanut butter at the same time over a low heat. Once you have a suitably gloopy mixture in the saucepan, add the other ingredients in turn, stirring frequently. When you are happy that the ingredients have been mixed together sufficiently, pour out into a baking tray to around 2½–5cm (1–2") in depth.

✦ Pop the baking tray into the freezer and cut out blocks from there as and when they are required.

Nutty Oatcakes

The final recipe of this trio – using the staples of suet, peanut butter and oats – is another fantastically simple recipe to put together and an ideal one to get the children involved with. Popular with almost every visitor to the garden, especially finches and thrushes.

Ingredients

1 cup animal suet or lard (you can use vegetable suet if you'd prefer)

1 cup crunchy peanut butter

2 cups porridge oats (preferably the quick-cooking microwave type)

2 cups cornmeal (or crushed stale cornflakes/rice crispies)

1 cup plain flour

1 cup wild bird seed

Method

✎ Melt the suet (or lard) along with the peanut butter over a low heat, in a largish saucepan. Once you are happy that they are fully mixed, add in the dry ingredients one at a time. When all the ingredients are thoroughly mixed, take a margarine tub and half fill it before freezing.

✎ If you want to re-use the tub after freezing, simply pop the frozen food block out of the tub, wrap it and put it back in the freezer.

✎ When using the blocks allow them to defrost before hanging them in the garden.

Pinecone Pleaser

Pinecones have a wonderful natural shape, which just cries out to be utilised as a bird feeder! This recipe is another quick, and very simple, way of providing nourishment for your garden birds. Next time you are out for a stroll, find the largest pinecones you can, and rustle up a pan of 'Pinecone Pleaser'.

Ingredients

4½ cups suet (animal or vegetable)

1 cup dried, crumbled bread (wholemeal is best)

½ cup shelled sunflower seeds

¼ cup millet seeds

¼ cup raisins (or, alternatively, chopped dried apples)

Pinecones (fully opened)

Method

✒ Melt your suet over a low heat in a saucepan. If you choose to use animal suet for this recipe, you may wish to render it first. Once the suet has melted, leave it to one side to cool down a little. While it is cooling, take a large mixing bowl and mix together the remaining ingredients (apart from the pine cones), stirring well.

✒ Once the cooling suet begins to thicken slightly, gradually stir it into the remaining ingredients, mixing them all thoroughly.

✒ Now for the pinecones! Stuff the mixture between the hard 'leaves' or 'petals' of the cone. Once the cones are full to the brim, hang them in the garden and enjoy the results.

Muffintastic!

It's those core suet and peanut butter ingredients again, but they prove

almost irresistible to so many species that come into the garden.

The suggestion is to hang them up in your bird feeders, but you can just

as easily use them by crumbling the muffins up and popping them out onto

the bird table. This is a variation on a theme, but a little variety in the recipe

and presentation department seems to be as appreciated

by the birds as it is by us!

Ingredients

1 cup suet (animal or vegetable)

1 cup peanut butter (smooth or crunchy)

3 cups porridge oats or cornmeal

½ cup wholemeal flour

Method

⚡ Put your cup of suet into a saucepan and melt slowly over a low heat. Gently stir in the peanut butter until well blended together. Once this is done, allow the suet/peanut butter mixture to cool until it begins to thicken. While this happens, blend the dry ingredients together and, once the suet/peanut butter has thickened, combine it with the dry ingredients.

⚡ Pour the mixture into a bun tray (or muffin tin, if you prefer) and freeze it. Use the buns as needed by hanging them in mesh food holders or putting them straight onto the bird table. As a gooey alternative, you could try and smear the non-frozen mixture on a tree trunk. You never know your luck!

⚡ The bun/muffin mixture can also be supplemented with raisins and breadcrumbs.

Black Treacle Balls

Part of the pleasure in cooking for birds is to see the results of your labours meet with a resounding 'thumbs up' from the garden visitors that you are catering for. Here is another 'bird balls' type of recipe that combines many familiar ingredients into a gooey sticky mess!

Ingredients

2 cups stale breadcrumbs

1 cup wild bird seed

¼ cup cornmeal (or crushed stale cornflakes/rice crispies)

1 cup peanuts (the ones you hang out in the red bag)

½ cup flour (wholewheat if possible)

½ cup grated cheese

½ cup sugar

¾ of a jar crunchy peanut butter

4 apples, chopped into small pieces

1 cup suet

1 cup raisins

1–2 spoonfuls black treacle

Method

✐ This is another recipe where all you have to do is mix everything in together! It's great fun when it's as easy as this!

✐ Mix up the following ingredients: breadcrumbs, cornmeal, flour, sugar, apples, raisins, wild bird seed, peanuts and cheese.

✐ Add the suet, peanut butter and black treacle. (The treacle is used to assist the binding process as much as anything else.)

✐ Once all the ingredients are well combined, you should be left with a fairly firm, yet still gooey, mixture. If the mixture is still a little on the dry side, you can add in a little more suet to bind them together.

✐ Shape this mixture into balls and leave them in the fridge to set. Place the chilled balls into a tough mesh bag and hang them up in trees, bushes or from the bird table. Additional balls can be frozen until you need them.

Further Reading

Golley, Mark, **The Complete Garden Bird Book**
(New Holland Publishers, 2001)

Golley, Mark, **Birdwatcher's Pocket Guide to Birds of Parks and Gardens**
(New Holland Publishers, 2004)

Hammond, Nicholas (Series Editor), **The Wildlife Trusts Guide to Birds**
(New Holland Publishers, 2002)

Moss, Stephen, **The Garden Bird Handbook**
(New Holland Publishers, 2003)

Moss, Stephen, **How to Birdwatch**
(New Holland Publishers, 2003)

Moss, Stephen and Cottridge, David, **Attracting Birds to Your Garden**
(New Holland Publishers, 2000)

Useful Addresses

The Wildlife Trusts
The Kiln, Waterside
Mather Road
Newark
Notts NG24 1WT
Tel: 0870 036 7711
Email: info@wildlife-trusts.cix.co.uk
www.wildlifetrusts.org

Wildlife Watch
Contact details as for The Wildlife Trusts
Email: watch@wildlife-trusts.cix.co.uk
www.wildlife-watch.org

British Trust for Ornithology (BTO)
The Nunnery
Thetford, Norfolk
IP24 2PU
Tel: 01842 750 050
Email: general@bto.org
www.bto.org

Royal Society for the Protection of Birds (RSPB)
The Lodge
Sandy
Bedfordshire
SG19 2DL
Tel: 01767 680 551
Email: bird@rspb.demon.co.uk
www.rspb.org.uk

Subbuteo Natural History Books Ltd
The Rea
Upton Magna
Shrewsbury
Shropshire
SY4 4UB
Tel: 0870 010 9700
Email: sales@subbooks.demon.co.uk

Index

Acknowledgements

Firstly, thanks to Jo Hemmings, Publishing Manager at New Holland, for giving me this rather unusual bird book project. Not what I'm used to, but enjoyable nonetheless.

Thanks to my Editor at New Holland, Gareth Jones, for his stoical patience and his easy-going badgering when the text was late, and to Rachel Lockwood for breathing life into the recipes through her artwork!

Finally thanks to my partner Nadine. This book's for you.

Artwork Acknowledgements

All artworks by Rachel Lockwood, with the exception of page 9 (Richard Allen).